1 MONTH OF
FREE
READING

at

www.ForgottenBooks.com

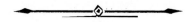

By purchasing this book you are eligible for one month membership to ForgottenBooks.com, giving you unlimited access to our entire collection of over 1,000,000 titles via our web site and mobile apps.

To claim your free month visit:
www.forgottenbooks.com/free1377570

ISBN 978-1-397-37217-8
PIBN 11377570

Forgotten Books is a registered trademark of FB &c Ltd.
Copyright © 2018 FB &c Ltd.
FB &c Ltd, Dalton House, 60 Windsor Avenue, London, SW19 2RR.
Company number 08720141. Registered in England and Wales.

For support please visit www.forgottenbooks.com

A STUDY OF THE CONDUCTIVITY OF CERTAIN ORGANIC

SALTS IN ABSOLUTE ETHYL ALCOHOL AT

15°, 25° AND 35°

by

ARTHUR M^{C}CAY PARDEE

DISSERTATION

Submitted to the Board of University Studies of

The Johns Hopkins University

at Baltimore

for the Degree of Doctor of Philosophy

1916

TABLE OF CONTENTS.

-:-   Page

ACKNOWLEDGMENT.

_ . _

The investigation under discussion in this dissertation was suggested by the late Professor H.C.Jones and was under his direct supervision up to the time of his death. The writer embraces this opportunity to pay a high tribute to the memory of this teacher who inspired his men with enthusiasm and led them by an example of tireless energy and close concentration towards a more thorough understanding of the fundamental truths of science. It is a pleasant duty to express an appreciation for the valuable instruction, inspiration and counsel which he has received both in the lecture room and in the laboratory from Professors Morse, Remsen and Swartz, and Associate Professors Lovelace, Frazer and Reid.

The author wishes to express his gratitude to Dr. H.H. Lloyd, his coworker, who has contributed much to the solution of the problems that have been encountered in the course of this investigation.

Dr. P.B.Davis has rendered valuable assistance in the designing and construction of apparatus necessary for this work, and for such the author feels his sense of obligation.

A STUDY OF THE CONDUCTIVITY OF CERTAIN ORGANIC SALTS IN ABSOLUTE ETHYL ALCOHOL AT 15°, 25° AND 35°.

INTRODUCTION.

-:-

For some years past a comprehensive study has been made in this laboratory of the electrical conductivity and dissociation of various organic acids in aqueous solution [1]. This naturally led to the investigation of the organic acids in absolute alcohol. Wightman, Wiesel and Jones [2], and Lloyd, Wiesel and Jones [3] undertook a thorough study of this latter work.

On account of their inability to obtain Λ_o for these acids experimentally, it was perfectly logical that the behavior of the organic salts in absolute alcohol be investigated. We have developed this for two reasons; first, we believe that accurate conductivity data on these compounds

---

1. Carnegie Inst. Wash. Pub. No. 170, Part II; No. 210, Chap. II.

2. Carnegie Inst. Wash. Pub. No. 210, Chap. III; J. Amer. Chem. Soc. 36, 2243 (1914).

3. Carnegie Inst. Wash. Pub. No. 230, Chap. VII; J. Amer. Chem. Soc. 38, 121 (1916).

4

is of value in the general consideration of the whole subject; secondly, we are interested in such minor questions
as temperature coefficients of conductivity, conductivity
in relation to chemical constitution, limits of experimental accuracy in working with dilute alcoholic solutions, and
the general phenomena of alcoholysis.

We realize also that here might be a method for obtaining the equivalent conductivity at infinite dilution,
(Λ_o), for the organic acids previously studied, by experimentally developing the Λ_o for the salts and from this
the migration velocities of the anions. Once obtaining
this data for the anions, it could be introduced into the
Kohlrausch equation for determining Λ_o theoretically, as
demonstrated possible for aqueous solutions by Ostwald [4].

With the values of Λ_o for both acids and salts, it is
of course possible to determine percentage dissociation
values as well as affinity constants for the acids in question.

HISTORICAL.

-:-

The phenomena of the electrical conductivity of the
sodium salts of organic acids in absolute alcohol up to the
present time have received but scant attention. With few

---

4. Zeit. physik. Chem. 2, 561 (1888); 3, 170 (1889);
   Amer. Chem. Journ. 46, 66 (1911).

exceptions all investigations that have been made were incidental in nature and were chosen simply as a type of the
organic salts.

Dutoit and Rappeport [5], in a study on the limiting
conductivities of some electrolytes in absolute alcohol,
measured among other salts sodium acetate, evidently taking
the same as an example of the salts of organic acids. They
subjected their results to some rather interesting deductions but unfortunately their conductivities were measured
at 18°, making exact comparison with those at 25° an impos-
sibility. They interpreted their results in a manner similar to that done by Goldschmidt, and so their deductions
are really illustrated in the latter's communications.

Dhar and Bhattacharyya [6] carried on some work in alcohol with various salts and studied among others the following organic derivatives: sodium propionate, sodium benzoate and sodium salicylate. Their measurements at odd
concentrations and temperatures render comparison impossible.

Heinrich Goldschmidt [7], incidental to his study of the
esterification of organic acids in absolute alcohol, found
it necessary to measure the conductivities of a number of
sodium salts of these acids. The preparation of the absolute

5. Jour. Chim. Phys. 6, 545 (1908).
6. Zeit. anorg. Chem. 82, 357 (1913).
7. Zeit. physik. Chem. 89, 129 (1914); 91, 46 (1916).

alcohol used by him is of interest. Ordinary 95 per cent.
alcohol was allowed to stand in contact with lime for some
time and then distilled. The process was repeated, this
time using a copper distilling vessel and condenser. Thus
the water content was reduced to 0.003 normal or 0.06 gr.
per liter. If completely anhydrous alcohol was desired,
this product was treated finally with metallic calcium.
For this purpose calcium bars were turned on a lathe to remove the coating of hydroxide, and the bright metal was cut
into pieces the size of a pea. An amount of calcium equal
to ten times the amount of water present (about 0.1 gr.)
was introduced into the alcohol. The whole was then heated
for several hours with a reflux condenser attached, and a
rapid stream of dry air was circulated through the distillation chamber to remove traces of ammonia. In this way
absolute alcohol was obtained having a specific conductivity
of 2×10^{-7}.

Having procured alcohol of sufficient purity, Goldschmidt
measured the conductivities of certain organic salts at 25°.
The salts were made, according to his bare statement, by
neutralizing the alcoholic solution of the acids with an
alcoholic solution of sodium ethylate made by dissolving
metallic sodium in absolute alcohol. Robertson and Acree [8]

8. Robertson, Dissertation, J. H. U., 1913.

describe accurately their method for preparing and stand-
ardizing this reagent in work which was carried out at a
much earlier date than this. Goldschmidt does not mention
the fact that he titrated his acids in alcohol against the
standard ethylate. The inference is that he standardized
the ethylate by titration with an aqueous solution of a
standard mineral acid using methyl orange as an indicator,
and then added arbitrarily the theoretical amount necessary
to a certain definite weight of the organic acid in alcoholic solution. He gives no experimental proof or statement as to whether he had tested the purity of the acids
used, Although it has been shown that such an assumption
is possible in the majority of cases, it is not always true
and we have shown it necessary to titrate the acid before
using the same. A further discussion of this is to be
found in the experimental part of this paper.

Goldschmidt measured the conductivities from N/10 to
N/5120 concentrations, and the values determined for five
different salts are shown below. These results are given
to enable us to discuss them and the deductions leading
from them, as well as to point out later wherein we differ
from him as to certain conclusions. These salts are sodium
trichloroacetate, dichloroacetate, picrate, salicylate and
sulfosalicylate. There is appended with each table his
calculation for Λ_0 of the salt at specified dilutions.

Table I.

Sodium Trichloroacetate.

V.	Λ I.	Λ II.
10	11.07	-
20	13.95	-
40	17.27	17.33
80	20.99	20.96
160	24.94	25.12
320	28.83	29.04
640	32.39	32.50
1280	35.28	35.29
2560	37.61	37.48
5120	39.23	38.92

$\Lambda_o^{(320-1280)} = 46.10$

$\Lambda_o^{(640-2560)} = 46.20$

$\Lambda_o^{(1280-5120)} = 45.52$

Mean $\Lambda_o = 4600$

Table II.

Sodium Dichloroacetate.

V	Λ I	Λ II	Λ III.
10	-	-	9.85
20	-	-	12.64
40	16.11	15.95	15.86
80	19.78	19.59	19.53
160	23.78	23.65	23.54
320	28.00	27.70	27.52
640	31.96	31.51	31.49
1280	35.66	34.87	34.96
2560	38.42	37.74	38.02
5120	-	40.71	40.86

$$\text{I} \begin{cases} \Lambda_o^{(320-1280)} = 48.54 \\ \Lambda_o^{(640-2560)} = 49\ 43 \end{cases}$$

$$\text{II} \begin{cases} \Lambda_o^{(320-1280)} = 47\ 68 \\ \Lambda_o^{(640-2560)} = 48\ 36 \\ \Lambda_o^{(1280-5120)} = 50\ 64 \end{cases}$$

$$\text{III} \begin{cases} \Lambda_o^{(320-1280)} = 47\ 36 \\ \Lambda_o^{(640-2560)} = 49\ 14 \\ \Lambda_o^{(1280-5120)} = 50\ 90 \end{cases}$$

Λ_o taken $= 52$

Table III.

Sodium Salicylate.

$V.$	$\Lambda.$
16	9.57
20	12.21
40	15.27
80	18.78
160	22.67
320	26.58
640	30.14
1280	33.20
2560	35.48
5120	36.29

$$\Lambda_o \,(320-1280) = 44.58$$

$$\Lambda_o \,(640-2560) = 44.7$$

$$\Lambda_o (1280-5120) = 41.55$$

$$-\text{Probable value} = 44.5$$

Table IV.

Sodium Sulfosalicylate.

$V.$	$\Lambda\, I$	$\Lambda\, II$	$\Lambda\, mean.$
40	13.56	13.54	13.5
80	16.72	16.74	16.7
160	20.21	20.18	20.2
320	23.76	23.69	23.7
640	27.06	27.02	27.0
1280	30.0	30.03	30.0
2560	32.22	32.23	32.2
5120	33.84	34.12	34.0

$$\Lambda_o(320-1280) = 40.7$$

$$\Lambda_o(640-2560) = 41.1$$

$$\Lambda_o(1280-5120) = 40.8$$

$$\Lambda_o = 40.9$$

Table V.

Sodium Picrate.

V	Λ I.	Λ II.
40	18.04	18.14
80	22.06	22.11
160	26.34	26.34
320	30.61	30.64
640	34.59	34.59
1280	37.94	38.07
2560	40.43	40.65
5120	42.03	42.75

$$I \begin{cases} \Lambda_o(320\text{-}1280) = 50\ 42 \\ \Lambda_o(640\text{-}2560) = 50.10 \\ \Lambda_o(1280\text{-}5120) = 48\ 99 \end{cases}$$

$$II \begin{cases} \Lambda_o(320\text{-}1280) = 50.72 \\ \Lambda_o(640\text{-}2560) = 50\ 97 \\ \Lambda_o(1280\text{-}5120) = 50\ 72 \end{cases}$$

Selected $\Lambda_o = 51.$

Goldschmidt thought that it was evident, after carrying his dilutions to 5120 liters, that Λ_o could not be reached by ordinary experimental methods. He attempted to calculate Λ_o for these organic salts and expected to obtain the relative velocity of the organic anion from the salt, and introduce the same into the equation -

$$\Lambda_o = \_c + a^{3}$$

To determine Λ_o for the organic salt he made use of the Kohlrausch formula [9] -

$$\Lambda_o = \Lambda + a\sqrt[3]{\frac{1}{v}}$$

in which Λ_o is an unknown, Λ the conductivity at a

---

9. Wied. Ann. <u>26</u>, 161 (1885).

known dilution, v, and a an unknown constant. Two equations involving the use of different Λ values are equated, the Λ_o being the same in both cases, and the expression solved for the value a. Once having this, it is a simple matter to solve for Λ_o in one of the two original equations. By reference to the tables quoted above we can observe how such values are derived. It is to be noticed that alternate Λ values are equated. This is done so that the difference may be of sufficient degree of magnitude and that any inaccuracy in an individual measurement may not affect two successive derivations.

A glance at the tables and calculations will show that the calculated values for Λ_o are by no means concordant. The higher the value of Λ used in the equation, the lower becomes the calculated Λ_o. His final conclusions are vague and must be regarded as only approximate and inconclusive. He generally chose the highest possible.

Goldschmidt seems to disregard entirely the very exact and admirable piece of work done on the subject of the limiting conductivity and degree of ionization of alcoholic solutions by B.B.Turner [10] in this laboratory. Turner carried his dilutions to far greater limits, as the following table illustrates.

---

10. Amer. Chem. Journ. 40, 558 (1908).

We have repeated this work and have every reason to believe that it is unquestioned and is remarkably accurate especially when one considers that it was done without the

Table VI.

KI in Absolute Alcohol
Conductivity in mhos at 25°.

V.	Λ.
10	22.2
12	23.0
16	24.1
32	27.5
64	31.1
128	35.0
250	38.2
500	41.4
1000	44.0
5000	47.8
10000	48.4
20000	48.5
∞	48.5 ± 0.5

more recent conductivity apparatus now at our disposal.

Turner showed that up to 5000 liters dilution it is easy to obtain concordant results; but the values for Λ_o

as calculated according to the Kohlrausch method are not
constant for these higher dilutions. Like those of Goldschmidt, they decrease the higher the values of Λ used in
the equation. Turner also showed that plotting Λ against
the reciprocal of the cube root of the volume does not give
a straight line as in aqueous solutions, of equal dilutions,
but rather a smooth curve slightly convex towards the dilution axis. He therefore assumed that the Kohlrausch method
fails to answer the requirements of absolute alcoholic solutions. Extrapolation of his results with the formula
would give a value of 56 for Λ_o instead of the experimental value of 48.5 obtained. He thought that accidental in-
troduction of water into his solutions might affect the
readings, so to test this he added as much as 0.2 to 0.3 %
of water by weight to his alcoholic solutions, with a variation in conductivity of only 0.01×10^{-6} units, showing
that no accidental experimental error of this nature had
crept in.

Furthermore, Dutoit and Rappeport [5] showed identically
the same phenomena with a number of inorganic salts in work
to which reference has already been made (page 6). This
work like that of Turner's seems to have escaped the notice
of Goldschmidt, as he does not mention either piece of work
in any of his papers.

In other words, the problem as undertaken by Goldschmidt

is very incomplete from this standpoint. No reason can be given why he should use arbitrarily chosen limits for v in applying the Kohlrausch formula, nor is it shown how accurately measured conductivities up to 20000 liters dilution can be reconciled with such a falling off in the calculated Λ_o for the salt.

Whether such a method could be applied or not, or whether another can be substituted in its place, is a question of very great importance. Furthermore, Goldschmidt based his conclusions on the results of only six or seven salts. It was therefore deemed advisable by the present writer that in the first place more conductivity data on a larger number of salts be obtained, and in the second place, these measurements be made at several temperatures in order to look at this subject in a broad way.

Such was the state of the problem at the inception of the experimental work on this dissertation.

EXPERIMENTAL.

-:-

Reagents.

The alcohol used in this investigation was prepared in the following manner. Ordinary 95 per cent. ethyl alcohol was heated for several days with lime in a copper tank with a glass condenser attached. A minimum of refluxing in the

condenser was obtained by inserting into the tank through
the stopper a coil of 3/16 inch lead pipe containing running water and serving to cause condensation immediately
below the reflux tube. The alcohol was distilled off using
a glass still head with a bulb blown in it and containing
glass wool soaked in alcohol in order to prevent any dust-
ing over of the dry calcium hydroxide. The middle fraction
was treated in the same manner as above and again fraction-
ated. This process was continued until a specific gravity
of 0.78507 was obtained, the extreme limits of variation
being 0.78505 to 0.78510, which according to Circular # 19
of the Bureau of Standards corresponds to a purity of from
100 to 99.987 per cent. The specific conductivities of the
alcohol varied with the different samples from 0.46 to 1.6
x 10^{-7} mhos. Upon the final distillation the alcohol was
collected in a six liter, alcohol extracted, Jena bottle
with a sealed stopper carrying a siphon for drawing off the
liquid, a calcium chloride - soda lime tube and an adapter
with a ground glass stopcock. Alcohol prepared and stored
in this manner, after several days following the distillation remained practically unchanged as to its conductivity
for a period of several weeks. It was found that our dis-
carded alcoholic solutions and washings, when distilled
once in glass with a few drops of concentrated sulfuric
acid before the final lime treatment, produced a very

superior grade of absolute alcohol, being generally better
than that obtained from fresh supplies of the 95 per cent.
material.

The organic salts used in this investigation were prepared by adding the necessary amount of sodium ethylate in
absolute alcohol to the organic acid in alcoholic solution,
as advised by Goldschmidt and previously mentioned in the
historical section (page 7). The acids employed were taken
from the various samples purified in the work of Lloyd,
Wiesel and Jones. When such were lacking new material was
obtained from well-known firms and purified in the following manner. Whenever possible the acid was precipitated
from hot alcoholic solution, but when necessary a small
amount of water was added. In every case the fractionation
was carried out several times. The halogen substituted
aliphatic acids were fractionally crystallized from hot
benzol, placed in a sulfuric acid dessicator, and the final
traces of the benzol removed by introducing into the con-
tainer pieces of paraffine which acted as an absorbant for
the solvent. To purify the liquid aliphatic acids we re-
sorted to both fractional crystallization by means of a re-
frigerant and repeated distillations under reduced pressure,
in the latter case collecting the various fractions in a
specially constructed receiver for small quantities.

The ethylate was prepared as needed in the following

manner, as suggested by J.H.Shrader [11]. A special grade
of metallic sodium, free from other metals, was wiped carefully with filter paper, the approximate amount was pared
to fresh surfaces, and in small pieces was placed first in
a good grade of alcohol, then transferred into some conductivity alcohol for final washing, and finally dropped
into a measuring flask of the best alcohol, so that upon
solution.it could be made up to the mark. With practice it
was possible to estimate successfully the amount of sodium
to produce a nearly N/10 solution. This solution was
standardized and used within an hour or two for the salt
preparation. It was found necessary to make the ethylate
to be used immediately, as evidences of decomposition giving a
straw color to the solution appeared within twenty four
hours of its preparation, and even sooner in the case of
more concentrated solutions.

This ethylate solution was immediately standardized
by means of a N/10 aqueous solution of hydrochloric acid.
This latter reagent was prepared by the method of Hulett
and Bonner [12], lately extended by Hendrixson [13]. As a
check on this solution four series of silver chloride grav-
imetric analyses were made at various times throughout the

11. J. H. Shrader, Dissertation, J. H. U., 1913, p. 14-16.

12. J. Amer. Chem. Soc. 31, 390 (1909).

13. J. Amer. Chem. Soc. 37, 2352 (1915).

year, none of which varied more than 0.1 of 1 per cent.

Phenolphthalein served as the indicator for the various titrations, special precautions - noted in a later paragraph - being used to prevent the interference of carbon dioxide from the atmosphere. As a final proof of the correctness of our choice of indicators the ethylate was standardized with hydrochloric acid using in this case methyl red as an indicator, and it showed results concordant with the phenolphthalein values previously obtained. The methyl red naturally was useless in the titration of most of the organic acids, so its use was abandoned after proving the value of the phenolphthalein procedure.

In order to dry completely our various pieces of apparatus, acetone was used, as suggested by Barnebey[14]. The acetone was dehydrated over calcium chloride and then redistilled.

Apparatus.

The cylindrical type of conductivity cells was used in all save the more concentrated solutions, where the ordinary plate type was adopted. The reason for using the cylindrical cell lies in the fact that the organic salts in absolute alcohol, although having more conductivity than the organic acids, are nevertheless of sufficient resistance

14. J. Amer. Chem. Soc. 37, 1835 (1915).

to warrant such a procedure. White [15] and Wightman [16] have described the method for obtaining the constants of these cells.

Both the temperature coefficients of expansion of alcohol and the temperature coefficients of conductivity of substances in it as a solvent are so large that it was especially necessary to maintain the solutions at a constant temperature to within 0.01°. The thermometers were of the differential Beckmann type and were carefully compared with a standard Reichsanstalt instrument which had in turn been calibrated at the Bureau of Standards. The gas regulator and thermoregulator combined was devised by Davis and Hughes [17]. The improved form of constant temperature bath as devised by Davis [18] was used in our investigation. These baths are capable of even finer temperature adjustment than that stated above as employed in our work.

The resistance box used throughout this work was calibrated at the Bureau of Standards. The improved Kohlrausch slide wire bridge was employed by means of which it was possible to read distances on the slide wire corresponding to tenths of a millimeter (the total length of the wire

15. Amer. Chem. Journ. 42, 527 (1909).

16. Amer. Chem. Journ. 44, 64 (1911).

17. Zeit. physik. Chem. 85, 519 (1913).

18. Carnegie Inst. Wash. Pub. No. 210, p. 21 (1914).

being five meters). Special precautions were taken to remove all external resistance in the circuit. Number 10
B. & S. insulated copper wire was used, with all leads com-
ing to the bridge dipping into a mercury contact rocking
commutator.

In the volumetric work Jena flasks were employed (50,
100, 200, 250, 500, 1000 c.c.) which had been previously
calibrated in this laboratory and recalibrated by ourselves
using weight methods. Reichsanstalt double mark pipettes
were recalibrated before use. In filling and draining the
pipette the following device was suggested by Dr. Davis.
It consisted of a right-angled T-tube with a glass stopcock on the base of the T, the pipette being attached by
rubber to one end of the cross piece held vertically with
the regulating finger on the opposite end of the cross
piece. The control finger is maintained throughout the operation at this opening and the danger of contamination by
suction is removed. A calcium chloride - soda lime tube is
inserted in the rubber tube leading from the glass stopcock on the base of the T to the mouth, for obvious reasons.
The 50 c.c. burettes adopted were calibrated at 2 c.c. in-
tervals by weight.

In order to titrate with phenolphthalein in an atmosphere free from carbon dioxide the following apparatus was
constructed partially as suggested by Hendrixson [13].

A carboy was connected to an ordinary tire pump and served
as a gas reservoir. The air was led through three wash
bottles, the first containing concentrated potassium hydrox_
ide solution, the second a more dilute solution and the
third pure water. The titration was effected in an Erlen_
meyer flask closed with a rubber stopper which in turn was
fitted loosely around the burette tip serving in this way
as a vent for the stream of air passed slowly through the
solution.

The difficulty in dessicating our acids when once purified was solved by means of a vacuum drying oven designed
by Dr. Davis and myself. This apparatus consisted of a
bell-jar 18 cm. by 24 cm. with a rubber stopper at the top
in which was inserted a thermometer and the evacuating connection. This tubulated ball-jar was placed on a heavy iron
vacuum plate and when in use the two parts were sealed by
means of rubber cement. Leading into the plate by means of
a rubber stopper from below were placed four wires; one
pair to a 110 volt 50 watt 16 candle power carbon filament
lamp placed in a metal chimney, and the other pair leading
to a miniature fan motor running on 110 volt direct current
with an eight candle power carbon filament lamp in series
with it outside the jar. The external electrical connec-
tions were made to enable the control of both heat and
power separately. Within the oven circulation was obtained

by means of the fan driving towards the open base of the
brass chimney. Drying was facilitated by means of two
dishes containing either concentrated sulfuric acid or
phosphorous pentoxide. The material was dried on watch
crystals placed on a perforated tray or shelf set above the
motor and lamp chimney. At 90 mm. pressure we were enabled
to get a boiling-point for water of 49.6°. Since the lamp
heating unit maintained a temperature of 65° it is easily
seen that with the added help of a strong dehydrating agent
all traces of the crystallizing solvent were removed. In
proof of this practically all the organic acids titrated
theoretical. This piece of apparatus has the following ad-
vantages: ease of construction, ability to be evacuated
sufficiently with the common water suction pump, and the
requirement of a lighting current for operation. Furthermore, it is convenient to have all parts of the apparatus
exposed to view through the transparent bell-jar, and to
open the oven it is only necessary to break the rubber
cement seal around the base of the jar.

Procedure.

The sodium ethylate prepared as previously described
was standardized by titration with N/10 HCl in a carbon dioxide freed atmosphere as described previously. When the
ethylate was standardized the organic acid from which the

salt was to be made was weighed out in quantity sufficient to give 100 c.c. N/10 salt solution and this weight was confirmed by titration, which showed a very general concordance giving added proof of the purity of the acids. In dealing with very deliquescent substances, as trichloracetic acid for example, we weighed by difference, making approximate standard solutions rather than exactly N/10 strengths; but even in this case we obtained confirmation of our work. The nondeliquescent, crystalline acids were weighed on a watch crystal, the deliquescent ones in glass stoppered weighing bottles; but in both cases the acids were washed through a funnel into the 100 c.c. measuring flasks with conductivity alcohol, and made up to mark at 25°. Several salts of N/50 dilution were made up in this same manner at the beginning of our work, but this dilution was omitted later as unnecessary.

Let us notice a few of the necessary steps in the titrations. All such were made in 70 c.c. solution.(50 c.c. water, 10 c.c. acid and approximately 10 c.c. ethylate). The carbon dioxide freed air was allowed to bubble through the solution for two minutes before titration. It was found that the presence of some alcohol retarded the endpoint and a number of titrations were made throughout the year to enable us to correct for this. We found as result of our work:

25

70 c.c.water &   0 c.c.alcohol required .03 c.c. to pro_
duce color.

60 c.c.water & 10 c.c.alcohol required .04 c.c.

50 c.c.water & 20 c.c.alcohol required .06 c.c.

Therefore it was necessary to apply this correction
as our accuracy in titration was made to check to .02 c.c.

After calculating the amounts necessary, 100 c.c. N/100
salt solution in absolute alcohol at 25° was prepared,
placed in a 150 c.c. glass stoppered Erlenmeyer flask, and
sealed with rubber cement until the conductivities were to
be determined. It was possible to make up three or four
different mother solutions of various organic salts in one
day, another day being devoted to the dilution down to
weaker concentrations, measurement of the conductivities
and calculation of results for each salt. These last three
operations on a single salt at various dilutions we have
designated as a "run".

It is deemed advisable at this point to introduce an
example of the calculations upon which a single salt was
prepared as described above.

Acid Orthonitrobenzoic $C_7H_5O_4N$.

Strength of standard HCl 0.10027.

I. Standardization of the Ethylate.

10.005 c.c. HCl used in each titration.

Ethylate Burette

Readings	Corrected	Difference
2.77 c.c.	2.76 c.c.	
10.84 c.c.	10.83 c.c.	8.07 c.c.
10.85 c.c.	10.84 c.c.	
18.91 c.c.	18.92 c.c.	8.08 c.c.
18.92 c.c.	18.93 c.c.	
26.99 c.c.	27.02 c.c.	8.09 c.c.

Mean 8.08 less .04 correction $\sim$ 8.04 c.c. ethylate.

10.005;8.04::x:.10027

x = .1248 normality of the ethylate

To make 100 c.c. N/100 salt solution requires 8.015 c.c.

II. Standardization of the Organic Acid.

10.005 c.c. acid used in each titration.

Ethylate Burette

Readings	Corrected	Difference
1.98 c.c.	1.98 c.c.	
10.03 c.c.	10.02 c.c.	8.04 c.c.
10.03 c.c.	10.02 c.c.	
18.05 c.c.	18.07 c.c.	8.05 c.c.
18.05 c.c.	18.07 c.c.	
26.08 c.c.	26.11 c.c.	8.04 c.c.

27

Mean 8.04 less .04 correction $=$ 8.00 c.c. ethylate.

$$8.04 : 8.00 :: .10027 : x$$

$$x = .09977 \text{ normality of organic acid}$$

To make 100 c.c. N/100 salt solution requires 10.02 c.c. plus .01 c.c. excess equals 10.03 c.c.

It should be mentioned that this work was carried on in a rather small room with one window and one door at opposite ends of the room, so that with care it was possible to keep the room temperature at 25° with less than 0.3° variation. Thus it was possible to measure out the solutions in burettes and pipettes provided that such were not handled unnecessarily to cause heating and were always kept dry to prevent cooling in evaporation. All burettes and pipettes were connected with calcium chloride - soda lime tubes to prevent contamination from moisture and carbon dioxide.

In handling the "run" the N/100 solution of one of the salts served as a basis for the preparation of all the more dilute solutions. The following scheme represents the method by which these solutions were prepared.

$$N/100$$

20 c.c. 50 c.c.	to soln.	20 c.c. 100 c.c.	to soln.	10 c.c. 100 c.c.	to soln	10 c.c. 200 c.c.	to soln.

$$N/250 \qquad N/500 \qquad N/1000 \qquad N/2000$$

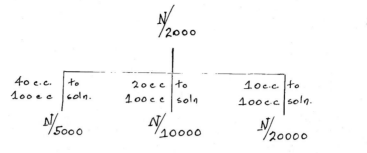

28

After a number of experiments it was deemed inadvis-
able to wash the measuring flasks with water; they were
therefore rinsed with a good grade of alcohol and then
three times with conductivity alcohol. The cells were
filled with conductivity water until several hours before
use. They were then rinsed three times with good alcohol.
Each cell was finally washed three times with the solution
of the particular dilution to be "run" in that cell before
filling. These cells together with one containing the conductivity alcohol were then introduced into the 15° bath,
gently agitated twice within an hour's time to insure absence of bubbles as well as to hasten diffusion, and then
read. They were placed successively in the 25° and 35°
baths allowing for the same time and procedure as in the
15° bath.

It will be remembered that the solutions were made up
at 25°, and that the molecular conductivities were measured
at 15°, 25° and 35°. Alcohol has such an appreciable temperature coefficient of expansion that it was necessary to
correct for the contraction and expansion at the other

temperatures. One liter of alcohol at 25° expands to
1.01114 liters at 35°, and contracts to 0.98923 liter at
15°. Therefore to obtain the molecular conductivity at 35°
one must multiply the specific conductivity at that temperature by the product of the molecular volume and the factor
1.01114. Likewise, to obtain the molecular conductivity at
15° the specific conductivity at that temperature must be
multiplied by the product of the molecular volume and the
factor 0.98923.

MEASUREMENTS.

-:-

Explanation of Tables.

In the following tables V signifies the volume at
which a solution was made up, and Λ the molecular conductivity of that solution at the various temperatures.
The method of calculating Λ is thoroughly familiar. Cor-
rections were applied as described allowing for the contraction and expansion of the solutions. (The solutions
were so dilute that their volume changes with variation in
temperature were assumed to be the same as that of pure alcohol). The values of Λ-25°, therefore, represent the
molecular conductivity of a solution of volume V at 25°.
The values of Λ-15° and Λ-35°, however, represent the
molecular conductivity of a solution of volume 0.98923 V

at 15° and 1.01114 V at 35°. Only the one value V is given
in the tables to save space.

All conductivities are expressed in reciprocal ohms.

Concerning the calculation of the temperature coefficients of conductivity, we have adopted this expression —

$$T = \frac{\Lambda \text{-} t' - \Lambda \text{-} t}{t' - t}$$

where $\Lambda \text{-} t'$ and $\Lambda \text{-} t$ represent the molecular conductivities of the same solution at $t'°$ and $t°$ $(t' > t)$, and T the
temperature coefficient of conductivity. To find the percentage coefficient of conductivity we have used the formula—

$$\Delta = \frac{T'}{\Lambda \text{-} t}$$

where Δ is the percentage coefficient and $\Lambda \text{-} t$ the conductivity at the lower temperature. At first the values of $\Lambda \text{-} t$
and $\Lambda \text{-} t'$ at 15° and 35° were corrected for the difference
in volume between 0.98923 V and V, and 1.01114 V and F,
respectively. This was done in order that comparison might
be made between solutions of the same volume. Later this
correction was omitted because of its small value.

Table VII.
Sodium Formate.

V.	Λ 25°	Λ 35°	Δ 25°-35°
100	20.09	22.70	1.30
250	25.03	28.53	1.40
500	28.48	33.25	1.67
1000	32.62	38.13	1.69
2000	35.34	41.88	1.85
5000	37.75	44.67	1.83
10000	39.02	46.35	1.88
20000	39.76	47.24	1.88

Table VIII.
Sodium Acetate.

V.	Λ 25°	Λ 35°	Δ 25°-35°
100	17.20	19.10	1.10
250	22.20	25.08	1.30
500	26.07	29.76	1.42
1000	29.99	34.67	1.56
2000	32.80	38.54	1.75
5000	35.42	41.62	1.75
10000	36.36	42.84	1.78
20000	36.79	42.95	1.67

Table IX.

Sodium Chloroacetate.

V.	Λ 15°	Λ 25°	Λ 35°	Δ 15°-25°	Δ 25°-35°
50	12.92	14.66	16.40	1.35	1.19
100	16.15	18.45	20.74	1.42	1.24
250	20.52	23.76	26.92	1.58	1.33
500	23.76	27.79	32.04	1.70	1.53
1000	26.51	31.34	36.56	1.82	1.67
2500	29.19	34.79	41.00	1.92	1.81
5000	30.73	36.81	43.83	1.98	1.91
10000	31.53	37.84	45.02	2.00	1.90

Table X.

Sodium Dichloroacetate.

V.	Λ 15°	Λ 25°	Λ 35°	Δ 15°-25°	Δ 25°-35°
100	18.12	20.84	23.67	1.50	1.36
250	22.38	26.06	29.91	1.64	1.48
500	25.55	30.00	34.92	1.74	1.64
1000	28.36	33.61	39.40	1.85	1.72
2000	30.30	36.12	42.79	1.92	1.85
5000	32.16	38.62	46.01	2.01	1.91
10000	33.08	39.77		2.02	-
20000	33.97	40.67	48.71	1.97	1.98

Table XI.
Sodium Trichloroacetate.

V.	Λ 15°	Λ 25°	Λ 35°	Δ 15°-25°	Δ 25°-35°
100	19.03	22.05	25.13	1.59	1.40
250	23.26	27.24	31.36	1.71	1.51
500	26.18	30.94	36.10	1.82	1.67
1000	28.80	34.31	40.31	1.91	1.75
2000	30.36	36.40	43.04	1.99	1.82
5000	32.24	38.79	46.12	2.03	1.89
10000	33.02	39.84	47.39	2.07	1.90
20000	33.71	40.54	48.39	2.03	1.94

Table XII.
Sodium Propionate.

V.	Λ 15°	Λ 25°	Λ 35°	Δ 15°-25°	Δ 25°-35°
100	14.68	16.50	18.18	1.24	1.02
250	18.80	21.41	23.95	1.39	1.19
500	22.16	25.71	28.95	1.60	1.29
1000	25.13	29.38	33.85	1.68	1.52
2000	27.35	32.28	37.66	1.80	1.67
5000	29.26	34.73	40.64	1.87	1.70
10000	30.27	36.11	42.52	1.93	1.77
20000	30.52	36.23	42.41	1.87	1.71

Table XIII.
Sodium Butyrate.

V.	Λ 15°	Λ 25°	Λ 35°	Δ 15°₂₅°	Δ 25°₃₅°
100	14.39	16.16	17.82	1.23	1.03
250	18.50	21.03.	23.49	1.37	1.17
500	21.79	24.97	28.37	1.46	1.36
1000	24.74	28.89	33.30	1.68	1.53
2000	26.93	31.80	37.07	1.81	1.66
5000	28.95	34.29	40.16	1.84	1.82
10000	29.85	35.67	42.06	1.95	1.79
20000	30.27	35.98	42.41	1.89	1.79

Table XIV.
Sodium Benzoate.

V.	Λ 15°	Λ 25°	Λ 35°	Δ 15°₂₅°	Δ 25°₃₅°
250	18.94	21.66	24.29	1.45	1.21
500	22.11	25.66	29.21	1.61	1.38
1000	25.03	29.40	33.93	1.75	1.54
2000	27.00	31.99	37.39	1.85	1.69
5000	29.18	34.78	40.83	1.92	1.74
10000	30.01	35.80	42.89	1.92	1.82
20000	30.41	36.18	42.64	1.90	1.79

Table XV.

Sodium Orthoamidobenzoate.

V.	Λ 15°	Λ 25°	Λ 35°	Δ 15°-25°	Δ 25°-35°
100	13.19	14.84	16.37	1.25	1.07
250	17.24	19.60	21.86	1.37	1.15
500	20.76	23.94	26.99	1.53	1.27
1000	23.90	27.82	31.92	1.64	1.47
2000	26.17	30.81	35.75	1.77	1.60
5000	28.44	33.54	38.91	1.79	1.60
10000	29.29	34.71	40.68	1.85	1.72
20000	29.52	34.82	40.50	1.80	1.63

Table XVI.

Sodium Orthochlorobenzoate.

V.	Λ 15°	Λ 25°	Λ 35°	Δ 15°-25°	Δ 25°-35°
100	14.54	16.49	18.34	1.34	1.12
250	18.71	21.52	24.25	1.50	1.27
500	21.90	25.44	29.09	1.62	1.43
1000	24.83	29.23	33.86	1.77	1.58
2000	26.85	31.96	37.42	1.90	1.71
5000	28.82	34.65	40.85	2.02	1.79
10000	29.97	36.04	..	2.03	–
20000	30.26	36.76	43.59	2.15	1.86

Table XVII.
Sodium Metachlorobenzoate.

V.	Λ 15°	Λ 25°	Λ 35°	Δ 15°-25°	Δ 25°-35°
100	15.53	17.69	19.82	1.39	1.20
250	19.74	22.76	25.80	1.53	1.34
500	22.83	26.62	30.59	1.66	1.49
1000	25.60	30.18	35.15	1.79	1.65
2000	27.38	32.68	38.36	1.94	1.74
5000	29.55	35.40	41.81	1.98	1.81
10000	30.51	36.49	43.40	1.96	1.89
20000	31.01	37.03	43.87	1.94	1.85

Table XVIII.
Sodium Salicylate.

V.	Λ 15°	Λ 25°	Λ 35°	Δ 15°-25°	Δ 25°-35°
50	14.19	16.32	18.42	1.51	1.29
100	17.19	19.87	22.58	1.56	1.36
250	21.53	25.13	28.84	1.67	1.48
500	24.56	28.92	33.60	1.78	1.62
1000	27.48	32.62	38.31	1.87	1.74
2000	29.18	34.87	41.18	1.95	1.81
2500	30.00	35.92	42.45	1.97	1.82
5000	31.20	37.47	44.57	2.01	1.87
10000	31.90	38.38	45.61	2.03	1.88
20000	32.38	38.99	46.36	2.04	1.89

37

Table XIX.

Sodium Acetylsalicylate.

v.	$\triangle$ 15°	$\triangle$ 25°	$\triangle$ 35°	$\triangle$ 15°-25°	$\triangle$ 25°-35°
50	14.16	16.29	18.38	1.50	1.28
100	17.16	19.81	22.50	1.54	1.36
250	21.50	25.08	28.71	1.67	1.45
500	24.44	28.73	33.41	1.76	1.63
1000	27.62	32.76	38.36	1.86	1.71
2500	30.06	35.89	42.38	1.94	1.81
5000	31.28	37.44	44.56	1.97	1.90
10000	32.56	38.97	46.45	1.97	1.92
25000	32.73	39.17	46.70	1.97	1.92

Table XX.

Sodium Sulfosalicylate.

v.	$\triangle$ 15°	$\triangle$ 25°	$\triangle$ 35°	$\triangle$ 15°-25°	$\triangle$ 25°-35°
50	12.38	14.28	16.21	1.53	1.35
100	15.30	17.73	20.29	1.59	1.44
250	19.19	22.48	25.92	1.71	1.53
500	21.92	25.90	30.22	1.82	1.67
1000	24.34	29.00	34.11	1.95	1.76
2000	26.12	31.27	37.11	1.97	1.87
2500	26.58	31.93	37.86	2.01	1.86
5000	28.18	33.88	40.42	2.02	1.93
10000	29.37	35.47	42.41	2.08	1.97
20000	30.65	37.13	44.16	2.11	1.89

38

Table XXI.

Sodium Orthonitrobenzoate.

V.	$\triangle$ 15°	$\triangle$ 25°	$\triangle$ 35°	$\triangle$ 15°-25°	$\angle$ 25°-35°
50	11.81	13.28	14.62	1.24	1.01
100	14.79	16.73	18.59	1.31	1.11
250	18.94	21.76	24.45	1.49	1.24
500	22.13	25.69	29.34	1.61	1.42
1000	24.87	29.26	33.82	1.77	1.56
2500	27.60	32.83	38.51	1.89	1.73
5000	29.10	34.89	41.29	1.99	1.83
10000	30.04	36.07	42.89	2.00	1.89

Table XXII.

Sodium Paranitrobenzoate.

V.	$\triangle$ 15°	$\triangle$ 25°	$\triangle$ 35°	$\triangle$ 15°-25°	$\triangle$ 25°-35°
50	14.31	16.47	18.61	1.51	1.30
100	17.22	19.96	22.72	1.59	1.38
250	21.04	−	28.19	−	−
500	23.98	28.36	33.09	1.83	1.67
1000	26.59	31.69	37.30	1.84	1.77
2500	28.63	34.25	40.51	1.96	1.83
5000	29.75	35.66	42.53	1.99	1.93
10000	30.70	36.98	44.10	2.05	1.93

39

Table XXIII.

Sodium 2,4,diñitrobenzoate.

V.	$\triangle$ 15°	$\triangle$ 25°	$\triangle$ 35°	$\angle$ 15°-25°	$\angle$ 25°-35°
100	18.20	21.09	24.09	1.59	1.42
250	22.18	25.98	29.93	1.71	1.52
500	24.95	29.45	34.39	1.80	1.69
1000	27.41	32.52	38.33	1.86	1.79
2000	28.74	34.42	40.88	1.98	1.88
5000	30.45	36.63	43.71	2.03	1.93
10000	31.21	37.66	44.94	2.07	1.93
20000	31.80	38.27	45.86	2.03	1.98

Table XXIV.

Sodium Orthotoluate.

V.	$\triangle$ 15°	$\triangle$ 25°	$\triangle$ 35°	$\angle$ 15°-25°	$\angle$ 25°-35°
50	11.42	12.80	14.11	1.22	1.01
100	14.19	16.03	17.81	1.30	1.11
250	18.32	21.07	23.67	1.50	1.23
500	21.28	24.69	28.16	1.60	1.41
1000	24.53	28.80	33.24	1.74	1.54
2500	27.17	32.19	37.52	1.85	1.66
5000	28.63	34.23	40.23	1.96	1.75
10000	29.56	35.36	41.75	1.96	1.81

40

Table XXV.
Sodium Paratoluate.

V.	$\triangle$ 15°	$\triangle$ 25°	$\triangle$ 35°	$\triangle$ 15°₋25°	$\triangle$ 25°₋35°
50	11.25	12.53	13.74	1.14	0.96
100	14.05	15.79	17.43	1.24	1.04
250	18.11	20.68	23.00	1.42	1.12
500	21.14	24.45	27.75	1.57	1.35
1000	24.15	28.29	32.55	1.71	1.51
2500	26.64	31.46	36.62	1.82	1.64
5000	28.00	33.29	39.21	1.89	1.78
10000	28.49	33.80	39.91	1.86	1.81

Table XXVI.
Sodium Picrate.

V.	$\triangle$ 15°	$\triangle$ 25°	$\triangle$ 35°	$\triangle$ 15°₋25°	$\triangle$ 35°₋35°
100	19.77	23.28	27.09	1.78	1.64
250	24.49	28.93	33.80	1.81	1.68
500	27.81	33.04	38.79	1.88	1.74
1000	30.65	36.56	43.24	1.93	1.83
2000	32.78	39.24	46.60	1.97	1.88
5000	34.73	41.67	49.85	2.00	1.96
10000	35.73	42.91	51.46	2.01	1.99
20000	36.32	43.63	52.43	2.01	2.02
40000		43.86			

41

DISCUSSION OF RESULTS.

The most apparent observation from the foregoing tab_
les is the great similarity in amount of conductivity of
these organic salts in alcohol. At 25° in a 1000 liter
dilution the extreme limits for the conductivity are from
26 to 36 mhos, with an average conductivity from 28 to 33.
The obvious reason for this is the uniform effect of the
sodium ion in the solution and the similarity in the velocities of the organic anions. As naturally expected, the
conductivities of these salts are much greater than those
of the acids.

Very little can be said as to the relation between
chemical composition and conductivity. The aliphatic and
aromatic derivatives show no difference, and the conductivity of the aromatic compounds seems to be independent of
the position of the various substituent groups. Sodium
picrate has a much larger conductivity than any other salt,
and the monosodiumsulfosalicylate at high dilutions gives
abnormally large and increasing conductivities, due prob-
ably to the secondary ionization of the carboxyl group at
these high dilutions.

In discussing the temperature coefficient of conductivity it is to be noticed that this value becomes gradually
larger with increase in dilution, and at the highest dilutions approximates the value 2.00. Just as in the

conductivity results there is here no definite relation between the values for the temperature coefficient and chemical composition.

It is of importance to note that this work on the sodium salts of the organic acids in absolute alcohol has been
greatly restricted owing to the almost complete insolubili-
ty of a great many of these salts in this solvent. If the
work were carried out in alcohol which was not absolute,
practically all the salts could be studied, for it is necessary to add a very small amount of water to obtain a sufficient degree of solubility. We have approximately cover-
ed the field of available compounds. It is of interest to
note that the polybasic acids of both the aliphatic and
aromatic series are excluded from study for this reason,
as well as all unsaturated acids of both series. A number
of salts of aromatic acids with di- and tri-substitutions
in the ring were likewise impossible for study.*

Reference has already been made (see pages 6 - 15) to
the work of Heinrich Goldschmidt on the conductivity of alcoholic solutions of sodium salts. We have purposely investigated most of the salts which he studied. A comparison

---

* It was the intention in this department previous to the
death of Professor Jones, to extend this work on the salts
into the field of mixed solvents (alcohol and water) since
a little preliminary work on our part has shown the possi-
bilities of this problem.

of these results is conveniently made by reference to the
following tables:

Salt	Goldschmidt	Author
Sodium Dichloroacetate	Table II p.9	Table X p.32
" Trichloroacetate	" I p.9	" XI p.33
" Salicylate	" III p.10	" XVIIIp.36
" Sulfosalicylate	" IV p.10	" XX p.37
" Picrate	" V p.11	" XXVI p.40

It can be seen from these tables that the two series
of conductivity values are in accordance, but an exact comparison cannot be made because of the fact that the values
of Λ in the two series refer to somewhat different concentrations. In order to make an effective comparison we have
plotted the values of Λ against the logarithms of the volume V in the case of sodium trichloroacetate (see Fig. 1).
The points circled refer to the data of Goldschmidt, and
the squares to that obtained in the present work. With few
exceptions all the points lie on one curve, and the slight
deviations which occur are within the limits of error of
the conductivity method. The four other salts give similar results; therefore their graphs are omitted.

It has been found impossible to obtain a limiting conductivity experimentally, although measurements have been
carried out to 10000 and 20000 liter dilutions. It is

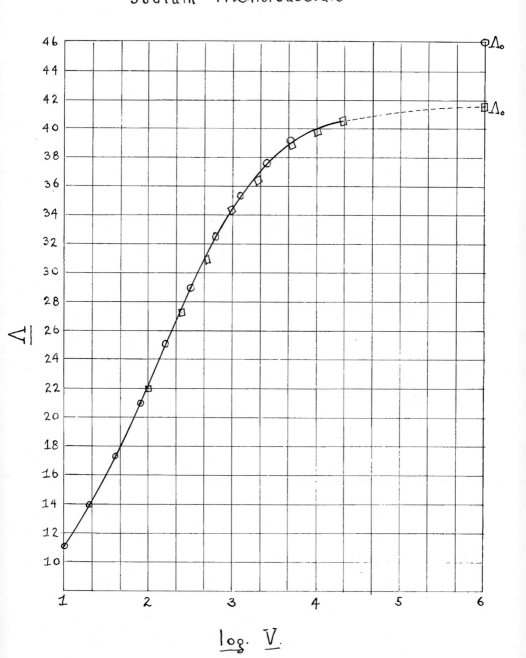

Fig. 1.

Sodium Trichloroacetate

therefore necessary to determine Λ_o by some method of extrapolation. It will be recalled that Goldschmidt used the
Kohlrausch formula for this purpose (see page 11), although
its applicability to alcoholic solutions and even to aqueous
solutions [19] had been previously questioned. We applied
this formula to our experimental data with a similarly un-
satisfactory result. The calculated values of Λ_o vary to
such an extent that it is impossible to make a selection.

A function of another form, suggested by A.A.Noyes [20],
which has been successfully used in connection with re-
searches upon the electrical conductivity of aqueous solutions, presented a possible means of determining Λ_o in
alcoholic solutions. This function has the form -

$$\frac{1}{\Lambda} = \frac{1}{\Lambda_o} + K \left(c\, \Lambda \right)^{n-1}$$

where Λ is the equivalent conductance at the concentration
c (1/V), K is a constant, and n is a number which, for
aqueous solutions, lies between 1.3 and 1.7. The value of
n is so chosen that the graph obtained by plotting the reciprocal of the equivalent conductance $(1/\Lambda)$ at the various concentrations (C) against $(c\Lambda)^{n-1}$ is nearly a straight
line. Two other graphs corresponding to neighboring values
of n, on opposite sides of the first line, are also drawn

19. A.A.Noyes, J. Amer. Chem. Soc. 30, 344 (1908).

20. J. Amer. Chem. Soc. 30, 335 (1908).

so as to aid in determining the most probable point at
which the graphs cut the $1/\Lambda$ axis [21]. This point is $1/\Lambda_0$,
the reciprocal of the limiting conductivity.

This procedure was followed, using the data at 25° of
sodium trichloroacetate, salicylate, orthonitrobenzoate,
and 2,4,dinitrobenzoate. The graphs obtained are in every
respect similar to those for aqueous solutions, except that
the value of n lies between 1.7 and 1.8. The values of Λ_0
obtained for the above salts at 25° are -

Sodium trichloroacetate	41.6
Sodium salicylate	39.9
Sodium orthonitrobenzoate	38.0
Sodium 2,4,dinitrobenzoate	39.2

From these figures the percentage dissociation of these
salts is obtained by means of the familiar formula - $\alpha = \dfrac{\Lambda}{\Lambda_0}$.

While the procedure outlined above is thus proved to
give satisfactory results in alcoholic solutions, the calculations are quite laborious, and advantage is taken of a
much shorter method of approximating Λ_0 , suggested by
Randall [22]. It is a fact that as the zero of concentration
(infinite dilution) is approached, the difference in the
percentage ionization of all salts approaches zero.

---

21. See J. Johnston, J. Amer. Chem. Soc. <u>31</u>, 1010 (1909).
22. J. Amer. Chem. Soc. <u>38</u>, 788 (1916).

47

Randall makes the provisional assumption that the ionization of salts of the same type (such as thallous chloride
and potassium chloride) is the same. Knowing the percentage dissociation of potassium chloride at various dilutions
very accurately, he calculates the value of Λ_o for thallous chloride by means of the equation -

$$\Lambda_o = \Lambda / \alpha'$$

in which α' is the percentage dissociation of KCl at any
given dilution, and Λ the molecular conductivity of TlCl
at the same dilution. Such a calculation gives values for
Λ_o which approach a constant figure with increasing dilution.

In applying this method to our results we have made
use of the values of percentage dissociation obtained by
means of the equation of Noyes. It has been found that the
four salts sodium trichloroacetate, salicylate, orthonitrobenzoate and 2,4,dinitrobenzoate include examples of all the
various types of salts encountered in the present investigation.

The calculated Λ_o is illustrated by the following
example:

V.	$100 \alpha_{25^\circ}$ Sod. salicylate	$\Lambda . 25^\circ$ Sod. acetate	$\dfrac{\Lambda-\text{Sod.acetate}}{\alpha-\text{Sod.salicylate}}$
100	49.8	17.20	34.5
250	63.0	22.20	35.2
500	72.5	26.97	35.9
1000	81.8	29.99	36.6
2000	87.4	32.80	37.5
5000	94.0	35.42	37.7
10000	96.3	36.36	37.8
20000	97.8	36.79	37.7

Probable $\Lambda_\circ$ = 37.7

49

The following table contains the probable values of Λ_o at 25° for all the salts studied by the author, calculated in the manner just indicated.

Sodium	Λ_o.	Referred to
Formate	40.7	Salicylate
Acetate	37.7	"
Chloroacetate	39.6	" & o-NO$_2$ benz.
Dichloroacetate	41.4	
Trichloroacetate	41.6	
Propionate	37.1	Salicylate
Butyrate	36.9	"
Benzoate	37.4	" & o-NO$_2$ benz.
Orthoamidobenzoate	36.6	Orthonitrobenzoate
Orthochlorobenz.	37.9	" & Salicylate
Metachlorobenz.	37.9	Salicylate
Salicylate	39.9	
Acetylsalicylate	39.9	Salicylate
Sulfosalicylate	--	
Orthonitrobenz.	38.0	
Paranitrobenz.	38.8	Orthonitrobenzoate
2,4,dinitrobenz.	39.2	–
Orthotoluate	37.3	Orthonitrobenzoate
Paratoluate	35.5	" & Salicylate
Picrate	44.7	Dinitrobenzoate

BIOGRAPHY.

-:

Arthur McCay Pardee, the author of this dissertation, was born in Rochester, New York, March 27, 1885. At an early age he removed with his parents to Pittsburgh, Pennsylvania, and received his preliminary training in the public schools of that city. After receiving his secondary education at the East Liberty Academy, Pittsburgh, Pennsylvania, he entered the freshman class of Washington and Jefferson College, Washington, Pennsylvania, in the fall of 1903. From this institution at graduation he received the degree of Bachelor of Arts in 1907.

During the year 1907-08 he served as Instructor in Chemistry at Tarkio College, Tarkio, Missouri, and in 1908 was appointed Professor of Chemistry at the same institution. In the fall of 1910 he entered The Johns Hopkins University as a graduate student in Chemistry. After a year's graduate work he returned to Tarkio College as Professor of Chemistry and Physics, and occupies that position at the present time. He was given a leave of absence during the year 1915-16 to enable him to extend his graduate studies. His subordinate subjects were Physical Chemistry and Geology.